HANDBOOK OF ENEMY AMMUNITION

PAMPHLET No. 2

GERMAN SHELLS, FUZES AND SMALL ARMS AMMUNITION

The Naval & Military Press Ltd

published in association with

Published by
The Naval & Military Press Ltd
Unit 10 Ridgewood Industrial Park,
Uckfield, East Sussex,
TN22 5QE England
Tel: +44 (0) 1825 749494
Fax: +44 (0) 1825 765701
www.naval-military-press.com

in association with

ROYAL
ARMOURIES

In reprinting in facsimile from the original, any imperfections are inevitably reproduced and the quality may fall short of modern type and cartographic standards.

CONTENTS TABLE

SHELLS

	PAGE
20 mm. Shell. Fig. 20	2
53 mm. Shell. Fig. 21	4
Skoda 37 mm. Armour Piercing High Explosive Shell with base fuze. Fig. 22	5

FUZES

Skoda base fuze. BZ 15–28–39. Fig. 23 ...	6
D.A. Fuze for 20 mm. Oerlikon A.A. Gun. Fig. 24	8

SMALL ARMS AMMUNITION

German Small Arms Ammunition. Fig. 25 ...	10

20 MM. SHELL.
Fig 20.

This shell, which is of unique design, was recovered without the fuze and consists of a solid drawn steel body with a hemispherical base. A driving band is fitted into a groove near the base. Immediately below the band, on the inside of the body, an aluminium ring is secured in a groove. The nose of the shell is fitted with a ring which is secured in the shell by four indents. The ring is threaded to take the fuze.

53 MM. SHELL.
Fig. 21.

The body of this shell, which is of cast steel, is in one piece and is threaded at the nose to take a direct action fuze the details of which are not complete. It is fitted near the base with a driving band and the exterior is not painted.

It has not yet been possible to identify the gun which fires this projectile as no gun of this calibre is known to exist in the German Army.. This projectile is, however, very similar to the 5 cm. shell Gn used with the 5 cm. gun on pedestal mounting K. i K as L or on A.F.B. 5 cm. K. i P.L.

SKODA 37 MM. ARMOUR PIERCING HIGH EXPLOSIVE SHELL WITH BASE FUZE.
Fig. 22.

This shell has a pointed and solid head having a radius of about 1½ calibres. The walls are tapered and threaded internally at the base to take a base percussion fuze. Externally the shell is fitted with a ballistic cap welded on and a groove is formed near the base to take a copper driving band; the groove is knurled to prevent the band turning on the shell. Immediately below the driving band a small groove is formed for crimping on the cartridge case to the shell. The explosive filling is Tolite contained in a cardboard wrapping.

The weight of the shell complete with fuze and gaine is 1·73 lb.

Note.—Pieces of 37 mm. nose fuzed shell have also been recovered. The dimensions and method of fixing to the case are the same as the above. No complete shell has, however, yet been found.

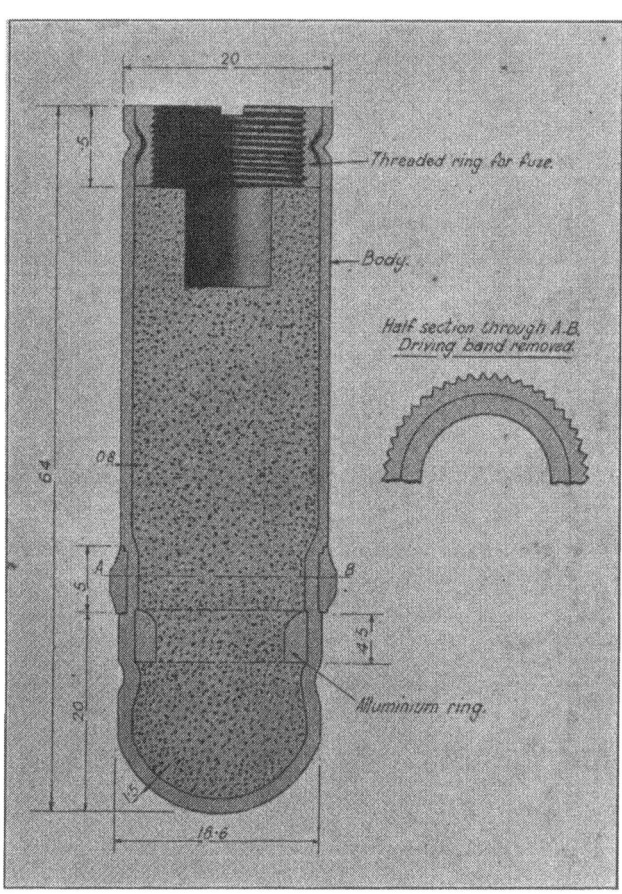

Fig. 20.

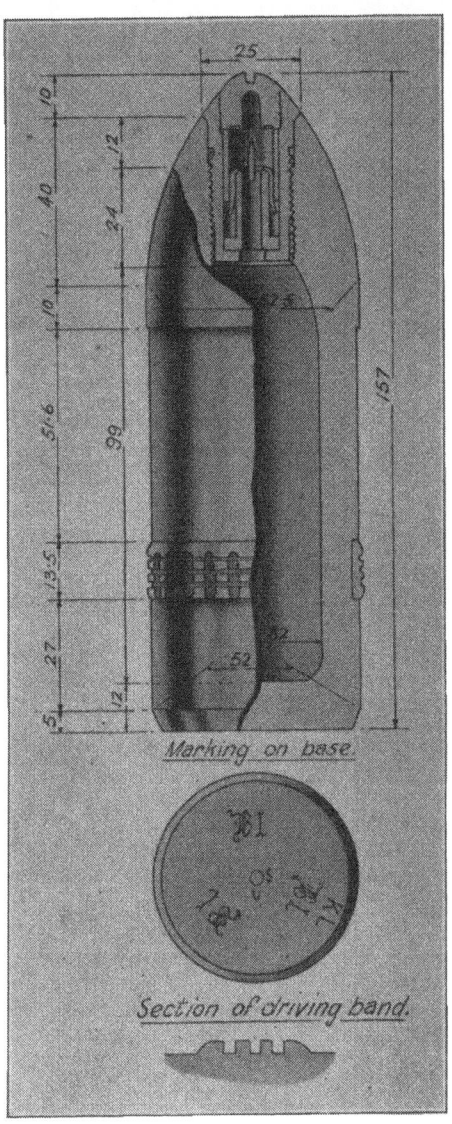

Fig. 21.

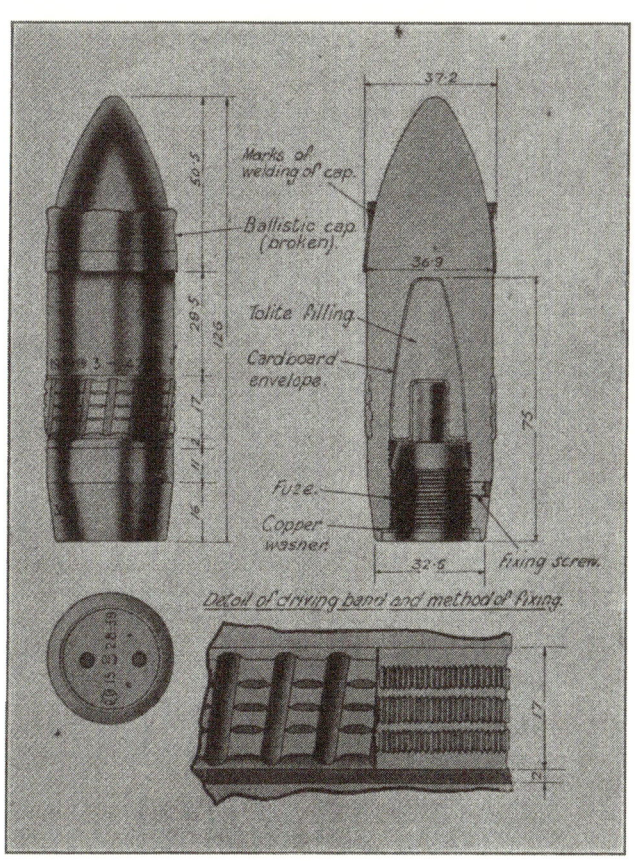

FIG. 22.

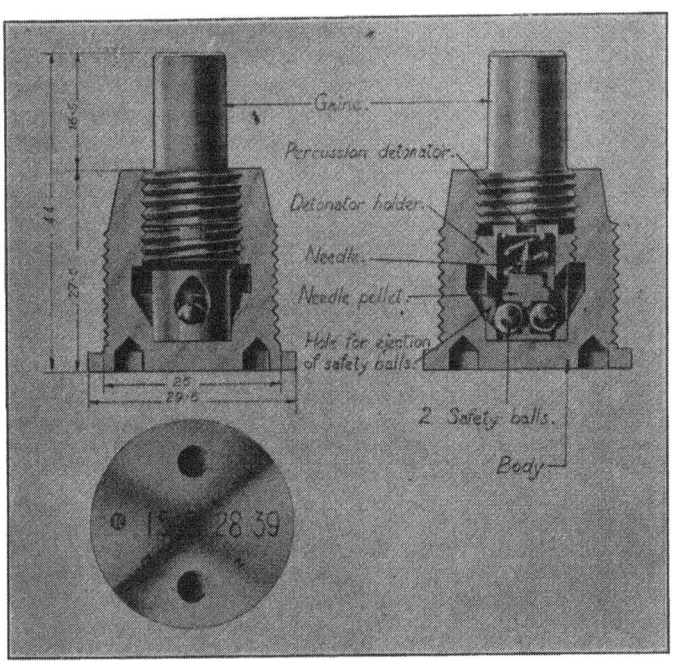

Skoda Base Fuze. BZ 15–28–39.

Fig. 23.

This fuze is screwed into the base of the shell on a copper washer and is secured by a fixing screw.

It consists of a body, detonator holder, needle pellet, creep spring and gaine.

The body is of steel, varnished black, threaded externally and formed with two recesses at the base to take a key for screwing into the base of the shell. Internally it is recessed and threaded to take the gaine and detonator holder. Two inclined grooves are cut in the body near the base, these lead into a lateral recess in the body into which the safety balls move after the shell has left the bore.

The aluminium needle pellet with steel needle is housed

inside the detonator holder. The pellet contains two polished nickel steel safety balls which are carried in a recess at the bottom of the pellet. The function of the safety balls together with the creep spring is to retain the pellet in a safe position before firing.

The gaine is of steel, varnished black. It contains the detonator and exploder and is screwed into the upper end of the fuze body against the detonator holder.

Action.—Before firing the needle pellet is kept from the detonator by the creep spring and the two safety balls. The latter, when the fuze is at rest lock the pellet to the detonator holder.

After firing the effect of set back keeps the pellet held against the fuze body.

On deceleration after the shell has left the bore the pellet tends to creep forward overcoming the resistance of the creep spring. This movement is assisted by the safety balls which, acting under centrifugal force, are caused to ride up the inclined planes in the detonator holder into the lateral recess in the body. The forward pressure of the safety balls on the pellet having ceased the creep spring reasserts itself and returns the pellet to its original position, this movement locks the safety balls in the recess in the body. The pellet is now held from the detonator only by the creep spring. On impact the pellet is carried forward on to the detonator overcoming the spring. The flash ignites the detonator in the gaine which in turn detonates the bursting charge of the shell.

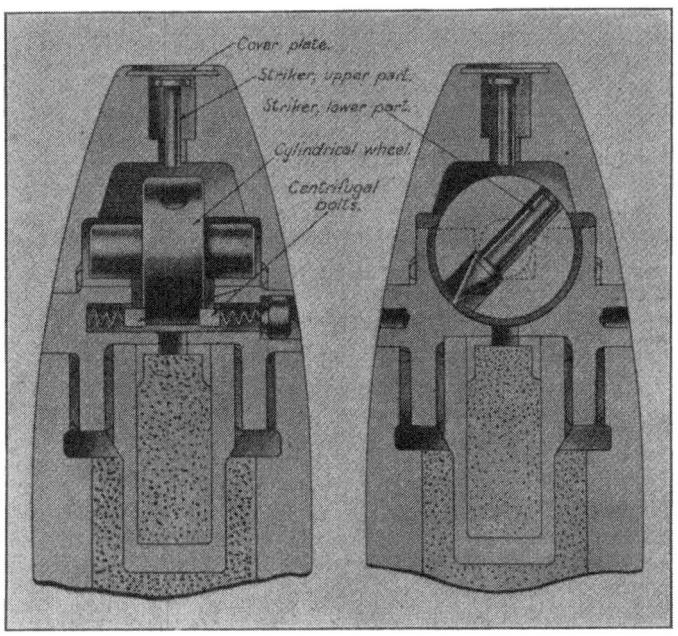

D.A. FUZE FOR 20 MM. OERLIKON A.A. GUN.

Fig. 24.

The striker is in two parts, the upper part being carried in a recess in the nose of the fuze ; the recess is closed by a cover plate. The lower part is housed in a cylindrical wheel carried in a recess in the fuze.

Before firing the various parts are assembled as shown in Fig. 24, the lower portion of the striker being inclined at an angle of about 45° to the longitudinal axis and retained in that position by two centrifugal bolts which fit in recesses one on each side of the wheel.

On firing the two bolts fly outwards under centrifugal force and free the wheel. The wheel, whose centre of gravity does not coincide with the axis of the fuze, tends, also under the action of centrifugal force, to revolve and

so bring the lower portion of the striker into line with the upper portion, but this movement is prevented by the effects of set back, due to acceleration in the bore, which forces the wheel back against the bottom walls of the recess in which it is placed.

On deceleration after the shell has left the bore, creep action causes the wheel to move slightly forward, it is then free to revolve and thus bring the two parts of the striker into alignment. Creep action and the protection of the cover plate against air pressure, keeps the striker from the detonator until it is forced in on impact.

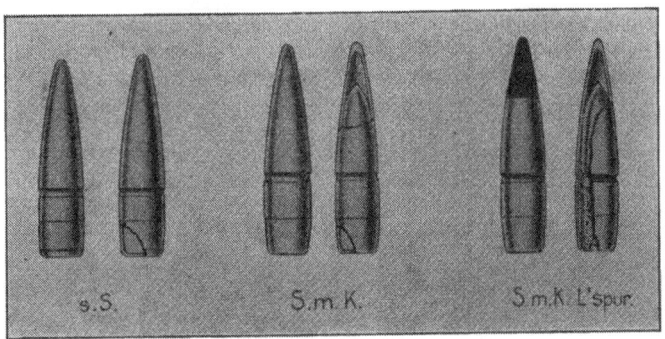

GERMAN SMALL ARMS AMMUNITION
Fig. 25.

Three types of German Small Arms bullets are shown in Fig. 25. These consist of a :—
  (a) s S cartridge.
  (b) S m K cartridge.
  (c) S m K L'spur cartridge.

A complete round consists of a cartridge case, percussion cap, propellant charge and bullet. The cartridge case may be either drawn from sheet brass (72 per cent. copper and 28 per cent. zinc) or from sheet steel, copper plated on both sides. The brass case is stamped S* on the base, the steel case with an " S " only.

The case is bottle shaped ; it is grooved at the base and coned slightly externally to facilitate extraction. A cap chamber is formed in the base of the case and connected by flash channels to the interior. In the centre of the chamber an anvil is formed on which the cap composition is fired by the striker.

The percussion cap may be either No. 88 or No. 30. The No. 88 consists of a brass detonator containing detonating composition and a covering cap of double sided zinc-plated lead foil. The detonating composition is put into the detonator dry and protected from damp and flaking by the

cap which is lacquered on the inside. The inside of the detonator is also lacquered to the level of the detonating composition.

The No. 30 cap is generally similar to the No. 88 differing in having certain components which, in the case of the No. 88 cause severe erosion, replaced by others without these disadvantages.

The base of the case is stamped with the Firm's mark, *e.g.*, P = Polte, the mark of the case, *e.g.*, S*, the delivery number, *e.g.*, 6 = delivery 6 and the year of manufacture, *e.g.*, 31 = 1931.

The propellant charge in the cartridges here described consists of nitrocellulose in blackish, square graphite treated flakes of about 0·25 mm. thick and 1·2 to 1·5 mm. long with smooth cut surfaces.

The bullet for the s S cartridge is of 7·9 mm. calibre and is formed with a groove at the base by means of which it is secured in the cartridge case. It consists of a bullet envelope into which the core is pressed. The envelope is drawn from ingot steel, plated with tombax and the core is pressed from hard lead. The base is streamlined. This bullet is the same for rifle, carbine or machine gun.

The S m K cartridge differs from the above in the bullet being somewhat longer. It contains a steel core around which there is a thin lead jacket. It is specially designed for armour piercing.

The S m K L'spur cartridge differs from the S m K in having a tracer bullet. The core is shorter than the S m K and has a case containing the tracer composition placed behind it. The composition burns green and red or yellow. The trajectory is marked by the burning of the tracer composition and can be seen up to 900 metres. It is used chiefly for A.A.

Marking.—To protect the detonating composition and the propellant from damp the annulus of the cap is lacquered. The colour of the lacquer is Green for s S cartridge and Red for S m K or S m K L'spur cartridge. In addition the point of the tracer bullet S m K L'spur is blackened for a distance of 10 mm. from the tip.

Packing.—The packing of service cartridges is as follows :

5 cartridges in one clip.
3 full clips in one folding box = 15 rds.
20 ,, folding boxes in one case = 300 rds.
5 ,, cases in one cartridge box = 1,500 rds.

A cartridge box filled with about 1,500 rounds weighs about 92·5 lb.

www.ingramcontent.com/pod-product-compliance
Lightning Source LLC
Chambersburg PA
CBHW032012080426
42735CB00007B/586